Stellar Thoughts

星思维

第九届“星艺杯”设计大赛获奖作品集

星艺装饰文化传媒中心　编著

广州新华出版发行集团
广州出版社

图书在版编目（C I P）数据

星　思维. 第九届“星艺杯”设计大赛获奖作品集：
汉文、英文 / 星艺装饰文化传媒中心编著. —广州：
广州出版社，2021.10
ISBN 978-7-5462-3360-4

Ⅰ. ①星…　Ⅱ. ①星…　Ⅲ. ①建筑设计—作品集—中
国—现代—汉、英　Ⅳ. ①TU206

中国版本图书馆 CIP 数据核字（2021）第 199103 号

书　　名	星　思维：第九届“星艺杯”设计大赛获奖作品集 Xing Siwei：Di-Jiu Jie “Xingyibei” Sheji Dasai Huojiang Zuopinji
编 著 者	星艺装饰文化传媒中心
出版发行	广州出版社 （地址：广州市天河区天润路 87 号 9、10 楼　邮政编码：510635 网址：www.gzcbs.com.cn）
责任编辑	刘雅丽　郑小燕
责任校对	王俊婕
排　　版	广州良弓广告有限公司
印刷单位	广州市赢彩彩印有限公司 （地址：广州市白云区嘉禾街鹤边鹤泰东路工业区 C 栋　邮政编码：510440）
规　　格	889 毫米 ×1194 毫米　1/12
印　　张	24.25
字　　数	30 千
版　　次	2021 年 10 月第 1 版
印　　次	2021 年 10 月第 1 次
书　　号	ISBN 978-7-5462-3360-4
定　　价	188.00 元

Create Happiness And Deliver Joy

设计幸福　播种快乐

星 艺 装 饰 文 化 传 媒 中 心

1 住宅·工程实景作品

2 住宅·方案设计作品

3 大宅·工程实景作品

4 大宅·方案设计作品

5 公共·工程实景作品

6 公共·方案设计作品

住宅·工程实景作品

RESIDENCE—ENGINEERING LIVE-SCENE WORKS

嘉裕公馆·马宅

MA RESIDENCE, JIAYU MANSION

项目名称：嘉裕公馆·马宅
项目设计：广东星艺装饰集团
设计师：谢雄峰

本案运用“以简胜繁”的设计理念，对原有的三房两卫空间进行合理规划，让视觉感受与使用功能得到完美结合。木制元素的运用彰显了高端、有质感的居家品位。简约的留白手法，巧妙地突出空间整体与局部的强烈对比，达到视觉的平衡。一把红蓝椅子的点缀，让整个空间简洁而不失大气，素雅又不失风采。

Following the design concept of "simplicity over complexity", this scenario redesigns the original three-bedroom and two-bathroom space so that visual experience and practical functions are perfectly combined. The wooden elements highlight the high-end and high-quality taste of living. The simple white-space technique cleverly underlines the sharp contrast between the whole and the part to achieve visual balance. The embellishment of a red-and-blue chair makes the whole space neat but magnificent, unadorned but elegant.

繁华之中的静谧空间

TRANQUILITY AMID HUSTLE AND BUSTLE

项目名称：繁华之中的静谧空间
项目设计：广东星艺装饰集团
设计师：丁捷

本案以白色为主调，整体风格力求极简，自然呈现材质最原始的美感。大面积采光区搭配开放的动线设计，形成绵延流畅的空间感；深灰色木板的设计加入，增加了空间的立体层次感，在光影和材质的互动中，营造出纯净、自然、宁静的空间氛围。

With white as the main tone, this scenario strives to achieve minimalism in overall style and to present the most primitive beauty of the materials. The large daylighting area together with the open circulation forms a continuous and fluid space. The addition of dark-grey wooden panels increases the three-dimensional layering of the space, creating a pure, natural and serene spatial atmosphere in the interaction of light, shadow and materials.

ARTS
PAPER SECRET

之·瘦

THINNESS

项目名称：之·瘦
项目设计：广东星艺装饰集团
设计师：刘天亮

本案希望居住者从外在走向内在，更加注重内心真正的需求。

设计师把“减负”这一生活认知，融入私宅的空间设计中，更注重空间与人的情感互动，希望人被生活温柔以待。

This scenario hopes that the residents can go beyond the appearance to attach greater importance to their inner needs.

By integrating the life philosophy of "burden reduction" into the design of the private home, the designer emphasizes the emotional interaction between space and people and hopes that people can be treated gently by life.

朔

SHUO

项目名称：朔
项目设计：广东星艺装饰集团
设计师：赵明明

本案是两套打通的全屋翻新，开放式厨房、早餐吧台及高配置的电器高柜都“藏”在方便使用的位置，放大了可敞开的空间，孩子可以自由地跑着、跳着，穿梭在家人的视线里，彼此独立又可以相互注意。本案在满足功能的同时兼顾装饰，空间重整划分后的布局，让每个人都拥有自己的舒适区，也让家人之间可以有更好的互动，提高了整体的生活品质。

This scenario is a refurbishment of two apartments made into one. The open kitchen, the breakfast bar, and the high cabinets of high-quality electrical appliances are all "hidden" in convenient locations, which enlarge the open space and allow children to run and jump freely within the sight of their family members. They are both independent of and attentive to each other. Both functions and embellishment are taken care of. The re-divided layout enables everyone to have their own comfort zone while maintaining better interactions between family members, thus improving the overall quality of life.

拾·光

GOOD OLD TIME

项目名称：拾·光
项目设计：广东星艺装饰集团
设计师：葛艺峰

家从来就不仅仅是一座房子，它承载着家人经历的一点一滴的美好时光。

本案业主进入大学前，一直和爸爸、妈妈、弟弟生活于此。老房子留下了全家人太多的温馨记忆。业主希望改造翻修后的老房子依然能够唤起当时的点滴，全家人在此延续与追寻更多的快乐时光。

Home is never just a house. It carries the good time experienced by the family.

The owner of this residence had been living here with her parents and younger brother until she went to university. The old house has left numerous warm memories of the family. By renovating the house, the owner hopes that the old house after renovation would recall the scenes of the past, and the family would extend and pursue more happy time in the house.

调

TONE

项目名称：调
项目设计：广东星艺装饰集团
设计师：魏歆

色彩是一门精妙的艺术。本案以灰、墨绿、橙为主色调，硬装造型、主材、选料、配色及软装饰品紧紧围绕配色规律来完成。既有“调调”，又不失“家”的舒适温暖！

Color is a delicate art. This scenario uses grey, dark green and orange as the main tone. The shapes, main materials, subsidiary materials and coloring of hard decoration, and the ornaments of soft decoration all follow closely the color matching rules, making the house both romantic and cozy.

领秀一方

LINGXIU YIFANG

项目名称：领秀一方
项目设计：广东星艺装饰集团
设计师：彭丽萍

本案特色在于造型设计：通过运用不同材质营造不同的空间氛围，从而体现出精致典雅的生活。

By emphasizing the design of shapes, this scenario creates different spatial atmosphere with different materials, thus embodying the delicate and elegant life.

嘉和城布洛可张府

ZHANG RESIDENCE, BLOCK, GENTLE TOWN

项目名称：嘉和城布洛可张府
项目设计：广东星艺装饰集团
设计师：陈忠波

设计之初，设计师问业主："你们想通过这次装修实现什么愿望？"

业主说："闲暇时间，掬一杯茶，捧一本书，伴着阳光慵懒地待着……"

美没有恒定标准，不是人人都能理解，但愉悦的感觉却人人都能感受到。好的设计就是让身居其中的人感到便利、舒适、愉悦。

At the very beginning, the designer asked the owner, "what do you want to achieve through this decoration?"

He answered, "In my leisure time, I can stay lazily in the sun with a cup of tea and a book..."

There is no constant standard of beauty. The sense of joy can be felt by everyone though not necessarily understood by all. A good design makes people inside feel convenient, comfortable and joyful.

乖乖的新房

GUAIGUAI'S NEW RESIDENCE

项目名称：乖乖的新房
项目设计：广东星艺装饰集团
设计师：张抗

本案设计师通过改变原始户型的空间比例和动线，增强了空间的功能性与流动性。灰白色系和秋香木色引领全屋调性，不同材质相互辉映，尽显空间质感。

The designer of this scenario enhances the functionality and fluidity of space by changing the spatial proportions and circulation of the original apartment. Off-white and dyed oak color dominates the whole residence. Different materials reinforce each other, highlighting the texture of space.

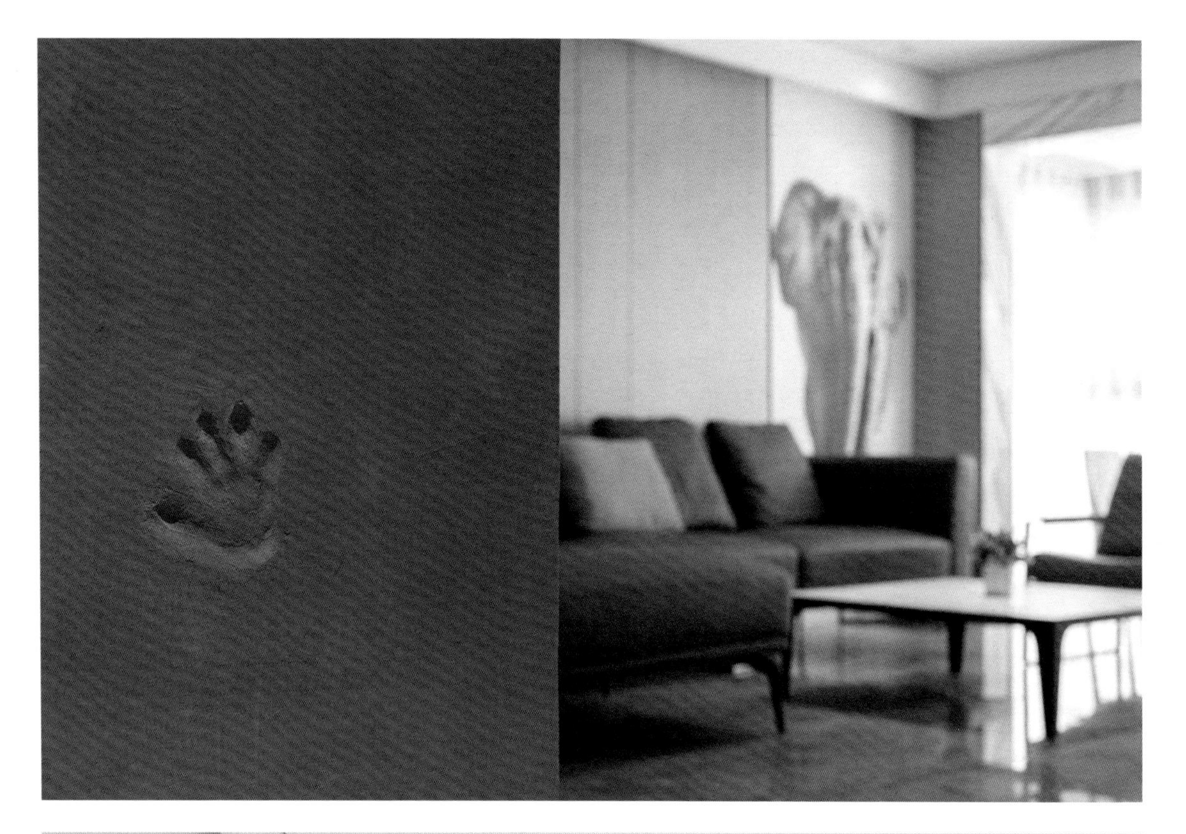

电视背景与入户背景相互延伸设计、客厅收纳柜悬空、大理石地台挑高配合暗藏灯带、秋香木沙发背景不规则分割拼接等精致的设计细节，在扩大整体空间感的同时，也提升了空间的温馨度与质感。卧房的床头背景为蓝色，木饰面打底，暗藏暖光灯带，营造温馨的睡眠氛围。

The exquisite design details, such as the mutual extension of the TV background and the entrance background, the suspended storage cabinets in the living room, the raised marble floor with the hidden light belt, and the irregularly divided and spliced dyed wood sofa, not only make the overall space seem larger but also improve its warmth and texture. In the bedroom, the blue background of the bedhead, the prevailing wood veneer, and the hidden light belt create a warm atmosphere for sleeping.

夏初

EARLY SUMMER

项目名称：夏初
项目设计：广东星艺装饰集团
设计师：刘洋

本案设计师通过设计拆掉原建筑中公共卫生间和客房的墙体，将整个空间释放出来，放大了客厅的横纵比例。次卫墙体的内移，扩大了生活阳台的面积。深棕色木材搭配爵士白石材墙面，配合浅灰色的瓷砖地面的设计，让空间更富层次感。顶面采用无主灯点式照明和条形照明，营造出整个空间的静谧。

The designer of this scenario removes the walls of the public toilet and the guestroom in the original structure to release the entire space to enlarge the length and width of the living room. The walls of the secondary toilet are moved inward to expand the area of the living balcony. The wall surface of dark brown wood with volakas stone, together with the pale grey tiled floor, make the space more layered. The no-main-light spot and strip lighting of the ceiling renders the whole space tranquil.

306

306

项目名称：306
项目设计：广东星艺装饰集团
设计师：赫晓春

本案设计师将原本的入户花园变成了集休闲区、鞋帽间、入户玄关为一体的多功能空间。大面积的高级灰与深、浅颜色的呼应，糅合出一种低调的高级感。开放式厨房的布局，既明亮又富有层次感。背景墙与极简的天花板设计，让空间在视觉上更有拉伸感。休闲区的设计使单一空间更富有生活气息。

The designer of this scenario changes the original entrance garden into a multifunctional space of leisure, cloakroom and hallway. The large areas of high-grade grey echoes the dark and light colors to create a low-key superiority. The open kitchen is both bright and layered. The background wall and the minimalist ceiling extends the space visually. The leisure zone makes the singular space full of life.

流光浮影

FLOWING LIGHT AND FLOATING SHADOW

项目名称：流光浮影
项目设计：广东星艺装饰集团
设计师：唐自立

在设计风格上定位现代轻奢，多点串合简约、简欧风格，满足居住者的不同需求。房子结构分明，有宾客行走动线、主人行走动线，以及工人行走范围。多采用度假、放松形式设计，在满足使用功能的基础上把别墅轻奢格调最大化。

The design style is modern light luxury. The combination of simplicity and simple European style satisfies the different needs of the residents. The whole residence is distinctively divided, with circulations for guests, owners and workers respectively. The design of vacation and relaxation is adopted to maximize the light luxury style of the villa on the basis of satisfying the practical functions.

朴舍

PUSHE

项目名称：朴舍
项目设计：广东星艺装饰集团
设计师：杨坚业

本案以咖色为主色调，采用简单大方的原木色，搭配形制简约的美式家具，体现了业主理性追求品质生活，不盲目、不空虚，彰显的是恰到好处的精致。

The main color of this scenario is brown, the simple and natural wood color, which matches the simple shape of the American furniture. The design represents the owner's rational pursuit of quality life, highlighting the perfect exquisiteness.

1001 何宅

1001 HE RESIDENCE

项目名称：1001 何宅
项目设计：广东星艺装饰集团
设计师：黄欢、何焱生

家是情感与爱交融的场所。本案力求打造出现代、雅致、温暖的居所空间。大理石背景墙的天然肌理与以自然纹理为主题的墙面，二者方位对立却相互呼应，营造出空间的默契，使得整体空间和谐而井然有序。

Home is a place of affections and love. The scenario tries to create a modern, elegant and cozy living space. The natural texture of the marble background wall and the natural patterns of the wall surface contrast but echo each other to create a harmonious and orderly space.

还

RECOVERING

项目名称：还
项目设计：广东星艺装饰集团
设计师：陈可可、莫琼

本案设计师根据居住者的互动模式、生活习惯，对空间结构关系进行梳理，对原结构进行重组，赋予了新的形式，以寻找和谐共生。新的空间秩序，让居住者有更好的体验和感觉，同时，设计师通过减少不必要的装饰元素，让朴素的东西发挥更大的原力。

The designer reorganizes the original structure according to the interaction mode and living habits of the residents to create new forms and seek harmony. The new spatial order gives the residents better experience and feelings. At the same time, the designer reduces unnecessary decorative elements to allow simple things to exert greater force.

新华御湖庄园

XINHUA ROYAL LAKE MANOR

项目名称：新华御湖庄园
项目设计：广东星艺装饰集团
设计师：唐乐云

本案设计让居住者在木质的纹理、温暖的阳光、清茶的幽香中，体验生活的安适感。主色调采用经典栗色，处处透露着对东方清雅生活意境的美好追求，整个空间中，没有任何显著的东方符号堆砌，言有尽，意无声，却自始至终浸透着东方特有的神思静谧。

The design allows residents to experience the comfort of life in the wood texture, the warm sunlight and the fragrant tea. The main color is the classic maroon, which reveals the pursuit of the elegant oriental life. Though without obvious piling up of oriental symbols in the entire space, it is fully imbued with the unique tranquility of the East.

住宅·方案设计作品

RESIDENCE—SCENARIO DESIGN WORKS

侨建御溪谷肖宅

XIAO RESIDENCE, ROYAL VALLEY

项目名称：侨建御溪谷肖宅
项目设计：广东星艺装饰集团
设计师：葛鑫

本案设计师运用教堂的采光方式，入口设计的三层中空，使整个空间更加开阔。窗口的特别设计，让空间在不同的时间产生不同的光影效果。同时，本案大面积使用灰色系的水泥涂料与原木相结合，置身其间有如伫立于教堂。二楼开放式的书房完美结合了中空空间与光影，给人一种简洁、质朴与大气的视觉感受。

The designer adopts church-style lighting. The three-storey hollow at the entrance makes the space more open. The special design of the windows creates different light and shadow effects at different time. At the same time, the large area of grey cement paint and the log make it feel like a church. The open study on the second floor perfectly combines the hollow space, light and shadow, giving people a visual experience of simplicity, naturalness and magnificence.

云·上

OVER THE CLOUD

项目名称：云·上
项目设计：广东星艺装饰集团
设计师：刘天亮

本案设计统括极简质感与视效，以高级灰和高端白为主色调，将功能与美学转化为精致的细节，配以简约精致的室内陈设，让整个空间弥漫着时尚又高级的气息。

This design is a unity of minimalist texture and visual effects, with high-end grey and white as the main colors. The functions and aesthetics are embodied in exquisite details. With simple and delicate interior furnishings, the entire space is filled with a stylish and high-end atmosphere.

魅

CHARISMA

项目名称：魅
项目设计：广东星艺装饰集团
设计师：程炜、王明健

本案设计将极简的生活艺术带进居室设计中。通过黑白的基调精心营造一处舒适的精神园地，以留白手法极大地唤醒人的想象力。阳光透过玻璃窗照进室内，光与影的互动，虚与实、白与黑、疏与密，错落有致，让有限的空间实现无限的可能。

The minimalist art of living is brought into the residence design. The black-and-white tone creates a comfortable spiritual garden while the white-space technique greatly awakens people's imagination. With the sunlight shining through the glass windows, light and shadow, virtuality and reality, black and white, sparseness and density are well arranged, enabling the limited space to realize infinite possibilities.

C House 陈宅

CHEN RESIDENCE, C HOUSE

项目名称：C House 陈宅
项目设计：广东星艺装饰集团
设计师：边诗琪

本案以大地色作为整体基调，内敛质朴但又不失大气，木饰面的温润与石材的精致碰撞融合，几何块面的交叠丰富了空间层次感。

This scenario takes the earth color as the overall tone, which is restrained but magnificent. The collision and integration between the warm wood veneer and the exquisite stone materials, and the overlapping geometric blocks enrich the sense of spatial layers.

半秋

MID-AUTUMN

项目名称：半秋
项目设计：广东星艺装饰集团
设计师：林朝杰

本案通过设计大面积的木饰面与白墙拼接，给人一种清新优雅、身心放松的惬意感。独立的大理石方柱是全屋的亮点，独特但不突兀，在家的每一个角落都能看到它特有的美。

In this scenario, the splice of a large area of wood veneer and white walls creates a refreshing, elegant and relaxing sense. The independent marble square column is the highlight of the whole residence. It is unique but not obtrusive, whose distinctive beauty can be seen from every corner of the house.

生长

GROWTH

项目名称：生长
项目设计：广东星艺装饰集团
设计师：冯劲皓

本案设计灵感来源于对生命的敬意，意在展示建筑融入生命、艺术融入生活。通过用绿植展示宇宙包罗万象的生机，从而营造一个有生命力的空间。

The design is inspired by respect for life. It aims at displaying that architecture is integrated into life and art into living. It creates a vital space by the use of green plants to show the all-encompassing universe.

意艺

MEANING AND ART

项目名称：意艺
项目设计：广东星艺装饰集团
设计师：徐潇

本案采用科技木、深海贝壳马赛克瓷砖、轻奢幽绿大理石，三种材料黄金比例分配，层次错落，低调优雅。宽阔的天花板采用功能性灯光布局，提升客厅的空间功能使用感。以纺织品、皮质来装饰的家具，给人一种低调有涵养的优雅感觉。开放式厨房被吧台分隔成两个区域，让人能自在地放松身心、愉悦用餐。每个空间都简单、实用而有气质，突显极简风格的本质。

This scenario uses scientific wood, deep-sea shell mosaic tile and light-luxury quiet green marble, with the three materials in golden ratio, making the residence well arranged, low-key and elegant. Functional lighting layout is adopted for the broad ceiling to enhance the sense of the spatial functions of the living room. The furniture decorated with textiles and leather renders the residence a low-key elegance. The open kitchen is separated into two areas by the bar, allowing people to enjoy their meals at ease. Every space is simple, practical and temperamental, highlighting the essence of a minimalist style.

锦绣·江南

SPLENDID JIANGNAN

项目名称：锦绣·江南
项目设计：广东星艺装饰集团
设计师：贺雷

本案设计整体空间精致、贵气、优雅。基于现代纯净的框架之上，空间格局收放自如，宁静、摩登、内敛，现代材质的组合再造了对新时代生活的想象。强大储物功能的设计，创造了整洁有序的空间视觉效果。

The overall space is exquisite, noble and graceful. Based on the pure modern framework, the spatial pattern is freely retractable, tranquil, modern and self-possessed. The combination of modern materials re-creates the imagination of life in the new era. The design of the strong storage function creates the visual effect of a tidy and orderly space.

卧室辅以温润柔和的自然木纹，营造出一个优雅温馨的空间。整体灯光布局让空间色调明亮、简洁且极具科技感，尽显居所的高端、品质、现代之感。

The bedrooms are decorated with gentle and soft natural wood grain to create an elegant and warm space. The overall lighting layout makes the space bright, neat and high-tech, rendering the residence high-end, high-quality and modern.

金碧华府

BABYLON LUXURY

项目名称：金碧华府
项目设计：广东星艺装饰集团
设计师：袁霄

设计师大胆打破空间格局，将楼梯移动到客厅区域与窗边的地台衔接，“岛式楼梯”环着木质的立面书架墙而上，构筑了一个协调连贯的开放式生活起居空间。双层挑高布局的空间和大幅玻璃引入了无边光景。人在空间中行走坐卧，目之所及皆是风景。全开放式的厨房与客厅连接形成了一个立体互动的家庭社交空间。

The designer boldly breaks the spatial layout by moving the staircase to the connection between the living room and the platform by the window. The "island staircase" along the wooden facade bookcase wall to construct a coordinated and coherent open space. The double-storey space and the large glass windows invite boundless scenery, which can be enjoyed by people either walking, sitting or lying. The fully open kitchen is connected to the living room to form a three-dimensional interactive social space for the family.

静之悠

TRANQUILITY

项目名称：静之悠
项目设计：广东星艺装饰集团
设计师：洪泉

本案巧妙地运用木色来提升空间气质，使整个空间看起来更加沉稳细致。一层空间木灰色沙发背景墙、简约的电视墙、温暖的壁炉都呈现出干净简单的视觉效果，也散发出独特的东方韵味。景观池营造出独特的意境，还保证了充足的通风和采光。主卧的顶面、床头背景运用了统一的木灰色墙板、麻布壁纸等材料，营造出一个简单、舒适的休息空间。精致细节的设计，兼具了现代装修的时尚性与实用性。

This scenario cleverly uses wood color to enhance the spatial temperament, making the entire space more stable and delicate. On the first floor, the wood grey color sofa background wall, the simple TV wall, the warm fireplace all display the clean and simple visual effect and exude a unique oriental charm. The landscape pool creates a unique artistic conception and guarantees sufficient ventilation and lighting. The ceiling and bedhead background of the master bedroom adopt uniform wood grey panel and linen wallpaper to create a simple and comfortable resting space. The exquisite details make the residence both fashionable and practicable.

300 ㎡轻奢狂想曲里的优雅舞姿

GRACEFUL DANCE IN 300 SQUARE METERS RHAPSODY OF LIGHT LUXURY

项目名称：300 ㎡轻奢狂想曲里的优雅舞姿
项目设计：广东星艺装饰集团
设计师：潘宇

本案通过门厅融合客、餐厅的整体化设计，让空间得到了极大的释放；电视背景与柜子合二为一，大大增加了收纳空间。金色元素与大理石的完美搭配，让衣帽间和卫生间尽显时尚轻奢感。微暖的乳白色将天花板和柜体相融合，营造出轻松的居住氛围。材质、色调的精心选择和运用，让各个功能空间既各自独立，又从整体上实现了空间的协调统一。

The integration of the hallway and the living and dining room greatly releases the space and the combination of the TV background and the cabinet into one increases the storage space. The perfect match of the golden elements and the marble make the cloakroom and the bathroom fashionable and luxurious. The slightly warm milky white blends the ceiling and the cabinet to create a delightful living atmosphere. The carefully selected and used materials and colors make different functional spaces independent of each other but coordinated and uniformed as a whole.

变形的珍珠

CHANGEABLE PEARL

项目名称：变形的珍珠
项目设计：广东星艺装饰集团
设计师：丁捷

本案将水泥的朴素之美与西方的装饰纹样相结合，展现了现代人于浮夸社会中寻求一丝净土的诉求。空间做旧的感觉如同在沙滩边捡到一个扇贝，轻轻打开壳面，一颗变形的珍珠透露出它的光芒，宣示着它的不凡，彰显着它的魅力。

The combination of the simple beauty of cement with western decorative patterns shows modern people's pursuit of a trace of pure land in this pompous world. The deliberately-made-old space feels like picking up a seashell on the seashore. When the shell is opened gently, a changeable pearl sheds its light, declaring its extraordinariness and highlighting its charm.

素阁

A PLAIN RESIDENCE

项目名称：素阁
项目设计：广东星艺装饰集团
设计师：张大为

本案以素雅稳重的木色叙述空间。强大的收纳功能设计，给空间营造安静、简洁、和谐的生活氛围。

Plain and steady wood color is used to describe the space. The design of a strong storage function creates a quiet, simple and harmonious living atmosphere.

晨光入舍

MORNING SUN IN THE RESIDENCE

项目名称：晨光入舍
项目设计：广东星艺装饰集团
设计师：徐诗钰

本案设计师大胆地对空间进行规整，通过简约的线条和纯粹的材料组合，采用利落的量体造型，充分展示了现代的美学生活。卧室中的飘窗被巧妙地改造成了小地台。客厅融入阳台的空间，更像是偷偷把大自然带进家里。当阳光从百叶窗中透入时，也照亮了居住者的心灵。

The designer boldly restructures the space by combining simple lines with pure materials, and by using neat tailor-made shapes to fully demonstrate the modern aesthetic life. The bay window in the bedroom is cleverly transformed into a small platform. The living room makes use of the space of the balcony, which is more like stealing the nature back home. When the sunlight penetrates through the blinds, it also illuminates the hearts of the residents.

流花 1401

LIUHUA 1401

项目名称：流花 1401
项目设计：广东星艺装饰集团
设计师：彭文博

本案用平淡色彩、素面材料营造安静的氛围，让窗外的自然湖景成为最好的装饰。设计师在阳台区置入半开放的活动区，保证了造型的完整性和通透性；客厅直出阳台的门，让动线更为多元化；北面房间的转轴门可以满足空间开放与私密的互换，让餐厅实现南北对流，光线也更加充足。

Plain color and materials are used to create a quiet atmosphere, turning the natural lake view outside the window into the best decoration. The designer puts a semi-open activity zone in the balcony to ensure integrity and permeability of the shape. The door of the living room to the balcony makes the circulations more diversified. The hinged door of the north room ensures the interchange of openness and privacy of the space and the north-south ventilation of the dining room while making the dining room brighter.

誉品原墅大平层

PENTHOUSE, ORIGIN VILLA

项目名称：誉品原墅大平层
项目设计：广东星艺装饰集团
设计师：陈汝顺

本案设计师通过改造大门获得独立门厅，让客餐厅通风透光；重新规划的餐厅与厨房比例也更协调。总空间 219 ㎡只保留了两个南套间，以便有更多的公共空间留给全家人交流。下班后，一家人聚在一起，欢声笑语充满屋内。如此的一个家，是温暖的、热情的、怀旧的、理性的，也是诗意的。

The designer renovates the gate to obtain an independent entrance hall and make the living and dining room better ventilated and lighted. The re-designed dining room and the kitchen are more proportionate. In the total space of 219 m^2, only two south suites are reserved, leaving more public space for communication among the family members. After getting home from work, the family gathers together, filling the home with laughter. Such a home is warm, enthusiastic, nostalgic, rational and poetic.

大宅·工程实景作品

MANSION—ENGINEERING LIVE-SCENE WORKS

唯·来

FUTURE

项目名称：唯·来
项目设计：广东星艺装饰集团
设计师：刘天亮

本案充分运用当代与复古的手法演绎空间，客厅与餐厅以素雅的焦磨黑和沉静的灰色作为硬装底色，高调配色与随心所欲的几何图案糅合兼并，再配以暖咖和玫瑰金不锈钢的陈设调和，深远的意境在设计中自然跃现。不规则的艺术装置与镜面的时尚组合、鲜艳的干枝花艺、错落的一字形灯饰，都以特有的艺术表现营造浪漫的氛围。

Contemporary and retro methods are fully utilized to interpret the space. The living room and the dining room use elegant burned black and calm grey as the background color of hard decoration. The high coloring scheme is combined with arbitrary geometric patterns, mediated by warm brown and rose gold furnishings to vividly show the profound artistic conception of the design. The fashionable combination of irregular artistic devices and mirrors, the floral art of bright dried flowers, and the strewn linear lighting all create a romantic atmosphere with unique artistic expressions.

Dentro di te

直线与弧线之间的空间对话

SPATIAL DIALOGUE BETWEEN STRAIGHT LINES AND ARCS

项目名称：直线与弧线之间的空间对话
项目设计：广东星艺装饰集团
设计师：姚国健

本案秉持“合适自然，合适生活，合适理想”的设计理念，将室内外空间一体化设计。整个空间运用凌厉直线和优雅弧线相结合，天花板上造型的四边直线和四转角弧线层层相叠、相隔展开；会客厅简洁的微弧电视背景墙来自知名汽车品牌宾利飞驰的溜背造型，搭配家具衬托空间设计，尽显空间的高端、大气、雅致。

With the design concept of "suitable for nature, for living and for ideal", this scenario applies integrated design for the indoor and outdoor space. The entire space combines sharp straight lines with elegant arcs. The straight lines on the four sides and the arcs at the four corners of the ceiling are overlapped. The simple and slightly arc TV background wall of the well-known Bentley flying back shape of the living room, matched with the furniture, fully demonstrates the high-end elegance of the space.

金地天玺私宅

PRIVATE RESIDENCE OF GEMDALE SIGNATURE

项目名称：金地天玺私宅
项目设计：广东星艺装饰集团
设计师：余文文

让极简融入设计，让设计融入生活，打破空间的界限感，让心与心更亲近。本案通过改动室内光线及墙面、家具等各个实物的设计和布局，营造出一种相对轻松自然的居住环境，从而给予居住者优质的精神体验。

Minimalism is integrated into design and design into life to break the spatial boundaries and to make hearts closer. By changing the design and layout of various objects such as indoor lighting, walls and furniture, this scenario creates a relaxing and natural living environment to offer the residents a high-quality spiritual experience.

“DU 品”——凡尔赛
“DU LUXURY”—VERSAILLES

项目名称：“DU 品”——凡尔赛
项目设计：广东星艺装饰集团
设计师：邹珊

本案采用法式传统略带复古的浪漫主义色彩，大胆运用色彩的冲突对比，独特的橙红色玄关和做旧白的客厅等，非常夺目。软装采用现代精简的视觉设计，突显时尚感。

Following the French traditional and slightly retro romanticism, this scenario boldly adopts the conflict and contrast of colors, such as orange hallway and made-old white living room, all very eye-catching. The soft decoration adopts a modern streamlined visual design to highlight the sense of fashion.

烙·克

BAROQUE

项目名称：烙·克
项目设计：广东星艺装饰集团
设计师：刘青峰

本案设计利用巴洛克风格的线条感与时代感来体现空间的气势与协调美，进而打造夸张、浪漫、充满激情的空间。整个设计打破均衡，形象多变，强调层次和深度。各色大理石、宝石、青铜装饰的使用，使住宅显得华丽且高贵。

The sense of lines and of time of the Baroque-style is used to represent the momentum and harmony of the space to create a spacious, romantic and passionate space. The entire design breaks the balance, with changeable images, emphasizing layers and depths. The use of various marble, gems and bronze decorations makes the residence gorgeous and noble.

沁翠

PENETRATING GREEN

项目名称：沁翠
项目设计：广东星艺装饰集团
设计师：魏歆

本案房体结构有足够的围度，却缺少了应有的高度。设计师大胆抹掉大宅里顶面和墙面的层次设计，避繁就简，同时配以色彩设计来保证大宅里的整体效果。

The house is big enough but not high enough. The designer boldly removes the layered design of the ceiling and walls to avoid complexity. At the same time, the color design guarantees the overall effects of the house.

胡宅

HU RESIDENCE

项目名称：胡宅
项目设计：广东星艺装饰集团
设计师：冷娟

本案不论是地面、壁面还是家具，木材都呈美式风格。家居常见的材质，尤其是洗白质感，不但能增添空间的明亮感，还能营造出岁月沉淀的空间氛围感。而木材温润的特性，则能为空间营造舒适的感觉。

The floor, the walls, the furniture and the wood materials are all of the American style. The commonly used materials for residencc, especially the bleached ones, not only make the space brighter, but also create a sense of history while the warm and gentle wooden materials offer a sense of comfort.

悠然假日

LEISURE VACATION

项目名称：悠然假日
项目设计：广东星艺装饰集团
设计师：魏己成

本案以休闲、度假为主题，采用现代简约的设计手法，呈现明亮、高挑、大气的空间结构，让人感觉舒适、恬静和放松。不仅注重居室的实用性，而且还体现出了现代社会生活的精致与个性，符合现代人的生活品位。

With the theme of leisure and vacation, this scenario adopts modern simplistic design techniques to present a bright, tall and magnificent spacc, offering people comfort, tranquility and relaxation. Not only the practicality of the residence but also the exquisiteness and individuality of modern social life are taken care of, which conforms to the taste of contemporary people.

丰顺别墅

FENGSHUN VILLA

项目名称：丰顺别墅
项目设计：广东星艺装饰集团
设计师：熊逸行

本案将现代元素和传统元素结合在一起，以现代人的审美需求来打造富有传统韵味的事物。居室的设计既保留了中式优雅、和谐的特点，又给人一种轻松明快的舒适感。

This scenario combines modern and traditional elements to create traditional charm based on the aesthetic needs of contemporary people. The design of the residence not only retains the characteristics of Chinese elegance and harmony but also gives people relaxation and comfort.

御景壹号

YUJING NO. 1

项目名称：御景壹号
项目设计：广东星艺装饰集团
设计师：谢锦盛

本案将经典奢华与北欧简约主义相结合，着力打造一个精睿低奢、富有人文艺术气息的品质居所。入户玄关的大块落地玻璃窗、左右两边的门洞造型，引领客人进入餐厅。客厅入口吊起的收纳柜，下面可放装饰品，丰富了空间的功能内涵。天花板斜边倒圆角再加拉丝条的勾勒，让空间层次丰富而有动感。简约的电视背景和镶入式的电视组合，配以深色的一字式电视柜，组成了极简的符号。

Classic luxury is combined with Nordic minimalism to create a high-quality residence with smart and low-key luxury and cultural and artistic flavor. The large floor-to-ceiling glass window of the entrance hall and two-door openings on both sides lead guests into the dining room. Under the hanging storage cabinet at the entrance of the living room, ornaments are placed to enrich the function of the space. The beveled fillet of the ceiling and the outline of wires make the space layered and dynamic. The simple TV background and the built-in TV, coupled with the dark linear TV cabinet, constitute a minimalist symbol.

保利溪湖

POLY RIVERSIDE LAKE

项目名称：保利溪湖
项目设计：广东星艺装饰集团
设计师：罗丹、金星燎

本案客厅挑空设计，端庄大气，以蓝孔雀图案作为沙发背景墙，与地毯花纹相呼应，打造出一种优雅精致的人文家居。色彩厚重的木质家具搭配圆滑灵动的造型，相得益彰，互为衬托。博古架的设计，让整个空间充满传统古韵，摆放的每一个物件，都是艺术品。大理石材的运用，丰富了空间层次，提升了空间质感。餐厅空间视觉感通透，无隔断设计使空间关系更紧密，生活场景丰富多彩。

In this scenario, the two-storey high living room is dignified and magnificent. The sofa background wall with the peacock pattern echoes that of the carpet, creating an elegant and exquisite home. The dark-colored wooden furniture and its smooth and smart shape complement and enhance each other. The antique shelf makes the entire space full of traditional charisma, every object on it being a piece of art. The use of marble enriches the spatial layers and improves the spatial quality. With no partitions, the dining room looks transparent, presenting closer spatial relationship and rich and colorful life scenes.

大宅·方案设计作品

MANSION—SCENARIO DESIGN WORKS

御景壹号高宅

GAO RESIDENCE, YUJING NO. 1

项目名称：御景壹号高宅
项目设计：广东星艺装饰集团
设计师：葛鑫

本案设计师通过柜子的摆放、材料的变换对空间进行不同的切换，让人在每一个空间中都能有不一样的感官体验。

The designer changes the space by altering the layout of the cabinets and the materials to offer people different sensual experience in every space.

房子用大面积的水泥涂料布局装饰，给人带来一种原始与安静的感觉。在局部设计上，运用木质材料为空间“升温”，让空间在水泥与木的结合中呈现出安静、原始和温馨之感。

The house is decorated with large areas of cement paint to bring about a primitive and quiet feeling. In individual parts, wooden materials are used to "heat up" the space so that people will have a quiet, original and warm feeling in the combination of cement and wood.

碧海之湾

BLUE SEA BAY

项目名称：碧海之湾
项目设计：广东星艺装饰集团
设计师：柳军友

本案通过设计扇形主阳台引景入室，让 180°城市景观一览无遗，无敌江景尽收眼底。天花板的无主灯设计清除了视野的遮挡，让城市的天际线得以完整保留。

The design of the fan-shaped main balcony allows people to have an unobstructed 180° view of the city and the river bay. The-no-main-light design of the ceiling clears all visual obstructions and keeps the city skyline intact.

无暗

NO DARKNESS

项目名称：无暗
项目设计：广东星艺装饰集团
设计师：郭剑阳、周兵

本案设计的出发点在于最大限度地让阳光和空气进屋，在满足功能需求的同时，能够让屋里的人跟阳光和景观产生联系。设计师通过合理规划动线，在负一楼设置了两个大尺寸的天井，再结合景观规划，让屋内的每个空间都能通风和采光。

The starting point of this design is to maximally let in sunlight and air. While satisfying the functional needs, it enables the residents to communicate with the sunlight and the landscape. By rationally designing the circulations, the designer sets up two large patios on the underground floor, and with landscape planning, every indoor space is ventilated and lighted.

艺术·家

ART HOME

项目名称：艺术·家
项目设计：广东星艺装饰集团
设计师：张大为

把生活融入艺术，让艺术藏入宅中。

本案通过融入园林小景等设计，结合现代的设计手法来渲染空间艺术氛围，该设计简约而不简单，以白色、木色为主，色调明亮简洁。G层为容纳社交和兴趣爱好的空间。一层为起居空间，两面庭院环绕。二层为居住空间。三层设立了书画室，马克·罗斯科的画在空间中点缀，增强了空间的艺术氛围，此空间设计同时衬托出了主人的优雅与内涵。

Let art be integrated into the life and hidden in residence.
By incorporating small garden landscape, this scenario uses modern design techniques to create the artistic atmosphere of space. The design is minimalist but not simple. Mainly white and wood color, the tone is bright and concise. The G floor is for socializing and hobbies. The first floor is the living space with courtyards on both sides. The second floor is the bedrooms. The third floor is the calligraphy and painting room, which is dotted with Mark Rothko's paintings to enhance its artistic atmosphere. The design of this space also shows the owner's elegance and upbringing.

卓越天元楼王顶复

KING TOP-FLOOR DUPLEX OF ZHUOYUE TIANYUAN

项目名称：卓越天元楼王顶复
项目设计：广东星艺装饰集团
设计师：范金如

本案业主是一个非常懂得追求生活品位的人，年轻有为，在接受这座城市的洗礼之后更加追求简约的生活方式，希望居住空间体现实用性的同时，还体现出现代生活的精致与个性。本案装饰材料以天然石材、木饰面、壁纸为主，营造出一个温馨、自然、舒适、健康的家庭环境。

The owner of this house is a promising young man who has excellent taste of life. After experiencing the ordeals of the city, he is keener on pursuing a simple lifestyle, hoping that the living space should not only be practicable, but also reflect the exquisiteness and individualism of modern life. The decoration materials are mainly stone, wood veneer and wallpaper, which create a warm, natural, comfortable and healthy family environment.

a hug in a mug.
GOOD
TARGET
Espresso
STUDY AND CONCEPT
START WITH
COFFEE
CREATIVE PROCESS

人与自然的探索

EXPLORATION OF MAN AND NATURE

项目名称：人与自然的探索
项目设计：广东星艺装饰集团
设计师：姚国健

本案设计师为了最大限度地把大自然的空气、阳光甚至雨水引进建筑空间，在户外庭院增加了凹形的“内廊”：人站在半开放的廊道上伸手就可以触摸大自然的一切。同时，通过设计让每个套房都光线充足，空气流通，转眼即景。

In order to let in as much natural air, sunlight and even rain as possible into the architectural space, the designer adds a concave "inner corridor" to the outdoor courtyard so that people standing in the semi-open corridor can reach out to touch everything in nature. At the same time, the design allows every suite to have sufficient sunlight and air and good view at the turn of the eye.

九号别墅

NO. 9 VILLA

项目名称：九号别墅
项目设计：广东星艺装饰集团
设计师：刘天亮

本案放弃形式化的设计符号，从情感深处挖掘意象，硬装的设计化繁为简：以纯粹的黑、白、灰色彩，通过简化视觉接触点，交汇铺陈出空间的基底，再运用不同材质、家具艺术品点缀，空间的背景无限后退，形成独特的空间氛围。居住者所有幻想、梦境、记忆都将成为空间的主宰。

In this scenario, formalized design symbols give way to images explored from the depth of emotions. The hard decoration aims at simplicity rather than complexity, with pure black, white and grey as the main colors. By simplifying the visual contact points, the spatial base is intersected and laid out. Dotted with furniture and artwork of different materials, the spatial background retreats infinitely, forming unique spatial atmosphere. All the fantasies, dreams and memories of the residents will become the dominators of the space.

沉光大境

GRANDVIEW OF TIME AND LIGHT

项目名称：沉光大境
项目设计：广东星艺装饰集团
设计师：谢洪

本案整体设计中，设计师将自然之景引入室内，空间因不同时间光线的穿透而流动，透过不同形态的墙体，让材料延伸，让空间不断放大。

In the overall design, the designer introduces the natural scenes indoors. The space flows with the penetration of sunlight at different time. The different forms of walls extend the materials and constantly enlarge the space.

时尚中灰别墅

FASHIONABLE GREY VILLA

项目名称：时尚中灰别墅
项目设计：广东星艺装饰集团
设计师：李先桂

极简主义是生活及艺术的一种风格，本意在于极力追求简约，并且拒绝违反这一风格的任何事物。本案根据业主喜欢黑、白、灰色调的气质，采用现代极简风格，整体简约整洁，突显优雅。

Minimalism is a style of life and art. It aims at pursuing simplicity and rejecting anything violating it. The owner of this scenario loves the tone of black, white and grey. The modern minimalist style makes the residence simplistic, neat and elegant.

凤凰郡起宅

QI RESIDENCE, PHOENIX TOWN

项目名称：凤凰郡起宅
项目设计：广东星艺装饰集团
设计师：廖春坤

“设计从心出发，从新出发”是本案的最终方向。“从心出发”，即去掉烦琐的制式，回归内心简单追求，让每一个细节都源自内心，一楼家庭成员交流、互动空间的设计，让家充满温馨，爱自然流动；“从新出发”，即注入时下生活的元素，用创新手法融合中式的设计和现代的生活。

The ultimate orientation of this scenario is "design starting from heart and from innovation". "Starting from heart" means to remove redundant format and to return to the inner pursuit of simplicity so that every detail comes from the heart. The communication and interaction space for the family members on the first floor makes the home warm and loving. "Starting from innovation" means to inject elements of contemporary life and to integrate Chinese-style design and modern life with innovative methods.

御金沙·刘宅

LIU RESIDENCE, DOLCE VITA

项目名称：御金沙·刘宅
项目设计：广东星艺装饰集团
设计师：谢雄峰

“少就是多，简洁就是丰富”。本案采用现代简约的设计手法，把室内装饰尽可能减少。同时兼顾功能设计实用性，讲究造型比例适度、空间结构明确美观、外观明快简洁，体现了现代生活快节奏、简约、实用又富有朝气的生活气息。

"Less is more and being concise is being rich". By adopting modern minimalist design techniques, this scenario tries to reduce the interior decorations as much as possible while stressing the practical functions, the proportionate shapes, the clear and beautiful spatial structure, and the bright and neat outer appearance. This residence reflects the fast-paced, simplistic, practical and modern life, which is full of vitality.

公共·工程实景作品

PUBLIC—ENGINEERING LIVE-SCENE WORKS

山水比德办公室

S.P.I. OFFICE

项目名称：山水比德办公室
项目设计：广东星艺装饰集团
设计师：谭立予

本案设计师将东向窗空间，设计成一个超过 20 米的前台，台面上是设计师亲手堆积的青苔山景。早晨阳光照进来，空间明亮又宽敞。前台与旁边三面墙体连接形成可以自动开合的会议室，设计师们可以在此集会。午餐时分，20 米前台变身长长的餐桌，设计师们还可以在此用餐、交流，一派温暖的生活气息。

The designer turns the east-facing window space into a front desk of over 20 meters. On the desk is the mountain view of moss piled up by the designer. The front desk is connected to the three walls to form a meeting room that can open and close automatically. It is a gathering place for the designers. At lunchtime, the 20-meter desk transforms into a long dining table for the designers to dine and communicate, which is a warm scene of life.

令合家日式烧肉

REIWA'S YAKINIKU

项目名称：令合家日式烧肉
项目设计：广东星艺装饰集团
设计师：冷金良

微黄的灯光，白色的灯笼，做旧纹理的原木饰面，融合到一起，完美地还原了日本动漫中的某些场景；一墙的日本酒瓶，各具特色的日本风情小海报，让食客们仿佛来到了日本街头小店……

The amber lamplight, the white lanterns, and the wood veneer with texture made old perfectly restore some scenes in Japanese animations. The whole wall of Japanese wine bottles and the unique Japanese-style small posters make diners feel as if coming to a small street shop in Japan.

亦小馆

YI CUISINE

项目名称：亦小馆
项目设计：广东星艺装饰集团
设计师：洪泉

本案空间中大面积运用中性色调，粗犷的砖红色、娴静的深芽绿、质感十足的金属色泽，烘托出整个空间的色调层次，质感间的统一、对比、和谐、跃动，都融于每一处细节中。点、线、面之间关系的灵活设计，让整体空间看似简洁直白，实则每一处细节都是由精雕细琢而来。

A neutral tone is applied to large areas in this scenario. The rough brick red, the demure deep bud green, and the full-textured metallic color all highlight the tonal layers of the entire space. The unity, contrast, harmony and dynamism of different textures are all blended in every detail. The flexible design of the relationship between dots, lines and planes makes the space seem concise and straightforward, but actually, every detail is carefully crafted.

滢·Beauty

Y·BEAUTY

项目名称：滢·Beauty
项目设计：广东星艺装饰集团
设计师：刘强

本案设计调整了门头立面比例，门头上方采用不锈钢百叶的形式，呈现出波浪般立体的视觉感受，搭配上灯源更显高端神秘。

The designer adjusts the facade ratio of the door head. The door head adopts the form of stainless-steel shutters, presenting a wave-like three-dimensional visual effect. Luminated by lights, it looks even more high-end and mysterious.

室内以米白色和高级浅灰色为主调，搭配墙面叠级的石膏线，提升了空间的层次感。大门右侧的来宾接待区，采用现代轻奢的元素，配合沙发组合设计营造出令人放松、舒适的空间氛围。中间隔断门洞采用不锈钢材质，在天花板点点光源的映衬下，尽显奢华质感。内厅私密的贵宾休闲区，通过轨道窗帘，搭配上金属屏风，营造出独特的“ins 风格”（即 Instagram 上流行的风格）。

The main tone of beige and high-class grey of the interior, coupled with the stacked plaster lines of the wall surface, enhances the sense of spatial layers. The guest reception area on the right side of the gate uses modern light luxury elements and a sofa set to create a relaxing and cozy spatial atmosphere. The middle partition door opening made of stainless steel shows a luxurious texture when illuminated by the dotted light source of the ceiling. The private VIP leisure area in the inner hall, with rail curtains and metal screens, creates a unique style popular on Instagram.

匀质空间

HOMOGENEOUS SPACE

项目名称：匀质空间
项目设计：广东星艺装饰集团
设计师：李飞

在重要的地方花费很多力气，可能反而将这个地方的美做得重复和过度。其实只要把次要的地方做好，重要的地方就会突显出来了，这就是空间的平衡感。在本案中，设计师致力通过块面、体量感穿插来表达理念，让工作者感受到空间的美感。

Too much emphasis on one important place may make its beauty repeated or excessive. Actually, so long as the secondary places are done well, the important ones will stand out. This is the sense of spatial balance. In this scenario, the designer is committed to expressing idea through the interspersed block surfaces and volumes so that workers can feel the spatial beauty.

大理古城那片云精品客栈

CLOUD HOTEL OF DALI OLD CITY

项目名称：大理古城那片云精品客栈
项目设计：广东星艺装饰集团
设计师：柯孟卿

本案巧妙运用几何线、面的交错和排列，舍弃突出物体，排除多余的装饰，秉承了日本传统美学中对原始形态的推崇风格。

By cleverly using the interlocking and arrangement of lines and planes, this scenario discards redundant decorations and adheres to the style of primitive forms in traditional Japanese aesthetics.

布达佩斯咖啡店

BUDAPEST CAFE

项目名称：布达佩斯咖啡店
项目设计：广东星艺装饰集团
设计师：黄柯

大理石台面、水磨石台阶、各种拱形和阶梯性的设计，都散发着“我很适合拍照”的信息。隐藏在拱形廊内的墨绿色长沙发靠墙延伸，大厅中间摆放着圆形桌，间距宽敞。裸露在外的结构和挑高的天花板、霓虹灯指示牌和透明泡泡椅等，各种精致的细节，让空间氛围清新雅致。

Marble-topped counter, terrazzo stairs, arched designs and mock staircases make the cafe an ideal place for taking photos. Hidden within archways, long dark-green banquettes extend against the wall. In the middle of the lounge, well-arranged round tables create a spacious place. Exquisite design details, including exposed structures, two-storey high ceilings, neon signs, transparent bubble chairs and so on, make the space fresh and elegant.

THE
budapest
CAFE

有梦想·有烧烤

DREAM · BARBECUE

项目名称：有梦想·有烧烤
项目设计：广东星艺装饰集团
设计师：贺雷

本案在设计墙面时整体加入吸音棉，巧妙处理了噪声扰民问题，同时，通过设计双层中空钢化玻璃，把光线引入室内，让空间更加敞亮、干净，又能隔绝用餐时的嘈杂。

Acoustic foams in the walls effectively prevent noise from disturbing the neighborhood. The double-glazed tempered glass brings light into the restaurant, making it brighter and cleaner, but also keeps noise in the restaurant from spilling out.

梦想烧烤
喝酒撸串
WiFi.
Mxskao
7508

TODAY
IS A
good
DAY
豪华型配餐柜

悟

CONTEMPLATION

项目名称：悟
项目设计：广东星艺装饰集团
设计师：彭淞

本案将中式元素与现代材质巧妙兼糅，唐宋风格的家具、布艺床品相互辉映，彰显出传统文化情怀，又给传统家居文化注入了新的气息。黑、白、灰主色调，用红色、铜色、绿植加以点缀，让整体设计格调雅致，造型简朴优美，色彩浓重而成熟，置身其中使人陷入深思、冥想。

This scenario is an ingenious blend of traditional Chinese elements and modern materials. Furniture in styles of the Tang and Song dynasties represents traditional culture, and fabric bedding adds some modern features to the whole. Mainly in black, white and grey, with red, bronze and some green plants as decorations, the overall design looks delicate. The simple and elegant shapes, and strong and mellow colors easily bring one into contemplation and meditation.

舒怡酒店

SHUYI HOTEL

项目名称：舒怡酒店
项目设计：广东星艺装饰集团
设计师：熊凌云

本案的设计为欧式风格，以纯净淡雅的黄色作为主色调，高品质家具、灯具以及艺术品的相互配合突出了酒店“雅”的文化氛围。灯光照明的设计不仅增强了空间的立体感与节奏感，还起到了连接酒店各个区域的桥梁作用。

This scenario is in European style, with pure and soft yellow as the main tone. The combination of high-grade furniture, lamps and other artworks manifests the elegance of the Hotel. The lighting design not only enhances the three-dimensional sense of the space and creates light strips full of rhythm, but also connects different areas of the hotel.

宅本全屋定制展厅

ZHAI BEN WHOLE HOUSE CUSTOMIZATION EXHIBITION HALL

项目名称：宅本全屋定制展厅
项目设计：广东星艺装饰集团
设计师：刘天亮

本案中，家的独特氛围，从入门一刻就能感受到，浅米色定下自然氛围基调，木质与皮革、石材融合，为家带来平静与温馨，富有韵律感的高光树瘤与金丝楠木拼接，如同一方开阔天地，盛纳万物，又仿若一处幽静园林，淡雅精致。

On entering the exhibition hall, one can sense its unique homey atmosphere. The natural tone set by light beige and the combination of wood, leather and stone create a peaceful and warm home. The rhythmic high-gloss burls spliced with silk wood make the hall look like an open and all-embracing space or a secluded and elegant garden.

MY WORLD

MY WORLD

项目名称：MY WORLD
项目设计：广东星艺装饰集团
设计师：丁雪

本案的设计定位为南美风格，元素源于精酿文化伴生出的雅皮文化，代表了一群反传统的年轻雅皮士的生活方式。设计师运用不同的高度差来划分功能区域，以增加空间的层次感。软装配以霓虹灯、圆形大舞台、阵列点灯、水磨石、木质地面、仙人掌等，以在复古和现代的结合中寻求质感上的突破。

In the South-American style, this scenario absorbs elements from yuppie culture, a concomitant of craft culture, to reflect the lifestyle of anti-traditional yuppies. Varied height differences divide the space into several functional areas, creating a layered look. The designer pursues an aesthetic breakthrough in the retro-modern mixture by decorations like neon lamps, a big round stage, lights in arrays, terrazzo, wooden floor and cacti.

PARTY
Beer world

青雅

CYAN ELEGANCE

项目名称：青雅
项目设计：广东星艺装饰集团
设计师：潘丽环、李年喜

本案设计以木制半透明的推拉门与墙面木装饰造型，配以冷静线条分割空间，抹去了一切繁杂与装饰。设计者运用简单的材料营造出豪华感，给人以既创新独特又大气优雅的环境体验。

This scenario abandons all redundant and complicated accessories, and uses translucent wooden sliding doors and wooden walls as decorations and simplified lines to divide the space. The designer uses simple materials to create a luxurious sense, providing a unique and elegant environment full of innovation.

硕

IMMENSITY

项目名称：硕
项目设计：广东星艺装饰集团
设计师：刘其

本案大胆采用个性化的混搭风。一桌、一椅、一空间，大片的光透过玻璃窗洒在室内灰色的地面上，置身其中仿佛与光、尘同行，听着空间述说它的故事，于是深思，并将它珍藏在记忆中。

This scenario boldly adopts a personalized mix-and-match style. A desk and a chair create a personal space. Through the window, a flood of sunlight is cast on the grey floor, making people feel as if floating with the light and dust. Immersed in such an atmosphere, they can listen to the space, reflect on its story, and cherish it in their memory.

荣县环海售楼部

RONG COUNTY HUANHAI SALES DEPARTMENT

项目名称：荣县环海售楼部
项目设计：广东星艺装饰集团
设计师：肖龙

原木＋大理石材的搭配，给人以自然质朴和舒适清新之感。前台大面积的木质背景，与顶面的格栅遥相呼应。玄关富有中国风的造型设计，与水池相衬相依，给人一种清幽又大气的感受。白色水墨纹路的大理石沙盘台与沙盘底部的灯带，配以展厅顶部的灯光设计，彰显出不一样的光影效果。洽谈区的木质地板和窗边的木质格栅，搭配布艺的沙发和略带轻奢风格的茶几布置，在淡雅自然的舒适感之外又多了一份正式、典雅。

The combination of logs and marble creates a natural, simple, comfortable and fresh feeling. The large wooden background behind the front desk perfectly coordinates with the grilles overhead. The hallway in Chinese style well matches the pool beside it, offering a quiet and magnificent feeling. Sand tables made of white marble with inky patterns, light strips beneath them, together with light design on the top, produce a unique effect of light and shadow. The wooden floor in the negotiation area and the wooden grilles by the windows, with fabric sofas and light luxury coffee tables, add formality and elegance to the overall simple, natural and comfortable atmosphere.

哈雷戴维森

HARLEY-DAVIDSON

项目名称：哈雷戴维森
项目设计：广东星艺装饰集团
设计师：刘浪

本案设计全景落地窗，将自然光照引入，旨在让空间回归自然舒适的状态。展厅前部分做了中空，进入展厅迎面是一整面醒目的红色标志，左边是前台咨询区，右边是展示区，顶面颜色是水泥深灰色，配合灯光让产品更出彩。后部分做了隔层，以便更合理地利用空间。

In this scenario, the panoramic floor-to-ceiling window brings natural light in, making a return to a natural and comfortable state. The front part of the exhibition hall is hollow. Facing the entrance of the exhibition hall is an eye-catching red LOGO occupying the whole wall. On the left is the front desk for consultation and on the right the display area. The cement top surface in dark grey and the lighting design make exhibits even more spectacular. In the back area, compartments are designed to make better use of the space.

公共·方案设计作品

PUBLIC—SCENARIO DESIGN WORKS

仁兰书屋

RELAX BOOKSTORE

项目名称：仁兰书屋
项目设计：广东星艺装饰集团
设计师：彭文博

"里仁为美，满径芳兰。"

"As long as there are inward virtues, fragrant orchids will bloom throughout the lane."

本案旨在打造让人爱上阅读、享受阅读的社区图书馆。在阅读区之外设置走秀 T 台及艺展区域，方便承办各种活动。“流云宅”可供从外地来广州的人士预约留宿，充分体现了“仁”的理念。

This scenario aims at building a community bookstore that will make people love and enjoy reading. Apart from the reading area, there is also a catwalk and an art exhibition area to facilitate various activities. "The Floating Cloud Residence" provides free accommodation for people from other places to Guangzhou if they have a reservation, which manifests the philosophy of benevolence.

空间设计保留了老建筑的框架，以“径”的意向为出发点，将各个独立的功能体块交错分布于小径两侧，如同一个没有花木的园林，光线透过工艺玻璃模糊了喧闹的外界，让人暂时逃离世俗纷扰，获得内心的宁静。

It preserves the framework of the old building and takes the "lane" as the core by arranging all the independent functional blocks on its both sides, making the space look like a garden without trees or flowers. Through the craft glass, lights from the room blur the bustling scenes outside, enabling people to temporarily escape from worldly noises and restoring in them a sense of tranquility.

砼馆

TONG CONCEPT

项目名称：砼馆
项目设计：广东星艺装饰集团
设计师：葛鑫

本案设计师将空间一分为二，体现了光与暗、疏与密、开敞与静谧。人们通过狭小的门洞在空间中不断转换，在两种冲突的环境中去体验水泥涂料所带来的安静与神秘。

The designer divides the space into two parts to reflect the contrast between lightness and darkness, between sparseness and density, as well as between openness and seclusion. Through the narrow doorways, people visit the two spaces back and forth and experience the tranquility and mystery brought about by cement paint in the two contrasting environments.

MYLANDS

天洋，迭代·序

TIANYANG, ITERATION · ORDER

项目名称：天洋，迭代·序
项目设计：广东星艺装饰集团
设计师：刘天亮

本案充分运用天花板与地面的延续轴线，配以片墙的阻挡形成回旋的空间。墙面造型的转折延伸，舒展视野。数组片墙捧起整个双斜翼天花板，连贯主题展台。大面积的通透墙体弱化了室内外的间隔，垂直与斜面以及交错穿射的光束赋予了三维空间里的多变性。

This scenario makes full use of the axes between the ceiling and the floor and builds blade walls as barriers, creating a revolving space. The turns and extension of the wall give people a broader horizon. Several groups of blade walls support the double-sloped ceiling, linking the thematic exhibition stand. Large areas of transparent walls weaken the sense of indoor-outdoor separation. Vertical structures, inclines and interlaced light beams make the three-dimensional space changeable.

怡宝
水处理专家

莲塘会所

LOTUS POND CLUB

项目名称：莲塘会所
项目设计：广东星艺装饰集团
设计师：李飞

本案设计出发点来自对自然光的利用和尊重。自然光既有永恒性也有多变性。人通过光来感知时间，设计师利用光塑造宁静氛围。

In this scenario, the design originates from the utilization of and respect for natural light which is both eternal and changeable. The designer uses light to create a quiet atmosphere.

四川花语精细化工有限公司总部办公室

HEADQUARTERS OFFICES OF SICHUAN FLOWER'S SONG FINE CHEMICAL CO., LTD.

项目名称：四川花语精细化工有限公司总部办
项目设计：广东星艺装饰集团
设计师：谢志兵

本案设计师以几何划分出个性的功能空间，加以蓝、白色的色彩碰撞，交织出活力、率性的氛围，让整个空间充满现代简约科技感。

In this scenario, geometric shapes divide the whole place into several personalized functional spaces. Together with the collision between blue and white, a vibrant and free atmosphere is created, filling the space with a sense of simplicity and technology.

大理卡蔓度假酒店

DALI KAMAN RESORT HOTEL

项目名称：大理卡蔓度假酒店
项目设计：广东星艺装饰集团
设计师：柯孟卿

本案设计贯穿了现代轻奢的手法，体现了实用性的同时也为传统的院落增添了几分时尚的气息。大面积的水磨石地面，暗藏地灯，给人以美观、大气的感觉。

Modern light luxury techniques are applied throughout the design, which makes the space practical and the traditional courtyard fashionable. Large areas of the terrazzo floor, with hidden floor lamps, create a sense of beauty and magnificence.

真美农霖整形

ZHENMEI NONGLIN ORTHOPEDIC HOSPITAL

项目名称：真美农霖整形
项目设计：广东星艺装饰集团
设计师：郭剑阳、彭丽萍

在本案设计中，设计师大胆将曲线融入室内设计，使空间产生一种行云流水般的艺术感。同时，整个项目的规划借鉴了琚宾先生的一些设计理念，即“游园平面”，设计师规划了一个弧形的游园平面，一步一景，步步是景，景在空间里，空间融入景中。

In this scenario, the designer boldly integrates curves into the interior design, creating an artistic sense of smoothness in the living space. Meanwhile, the overall plan draws on some design concepts of the Chinese architect Ju Bin, such as "garden tour surface". With the cambered garden-tour surface, every step leads to a different scene, because the scenery and space are rolled into a whole.

十里

SHI LI

项目名称：十里
项目设计：广东星艺装饰集团
设计师：刘永明

本案用极简解构的手法把空间打散重组，在满足实用功能的基础上，让空间既有区域分割感，又是呼应的一个整体。地面、天花板、墙面虽然都是独立的体块，但是共同创造出了一个完整的空间。一层的双向楼梯设计，就像一尊极具视觉冲击力的雕塑矗立在展台上。叠级的台阶，配上人体感应的灯光线条，彰显设计感。整体极简的设计让空间更显简洁、大气。

Guided by minimalism and deconstructionism, the designer of this scenario breaks the space into several parts and reorganizes them, which not only fulfills practical functions but also creates a sense of division in the whole space. The floor, the ceiling and the wall, although being independent blocks themselves, together create a complete space. The two-way staircases on the first floor look like a statue with great visual impact standing on the stage. The stacked steps and the human body induction light strips are highly fashionable. Minimalism in the overall design makes the space neater and more magnificent.

西影私宅·归园田居

XIYING PRIVATE RESIDENCE—A RETURN TO NATURE

项目名称：西影私宅·归园田居
项目设计：广东星艺装饰集团
设计师：李玉明

本案庭院包含了餐厅、展览空间、手工艺作坊、烹饪工作室、洽谈工作室和儿童游乐场等一系列公共空间。客人们可以在此体验民族特色，或者在周围的山间探索游玩。

In this scenario, the courtyard includes various public spaces, including a restaurant, a display area, a handicraft workshop, a cooking studio, a negotiation studio and a children's playground. Guests can experience distinctive features of ethnic minorities or explore and play in the nearby hills.

南涧售楼部

NANJIAN SALES DEPARTMENT

项目名称：南涧售楼部
项目设计：广东星艺装饰集团
设计师：林中彬

本案采用简约流动的设计手法，在材质选择上力求克制，通过空间的层次配合，自然而然地表达现代美学品位。售楼部正中深浅搭配的大理石沙盘区，配以顶部序列化的挑空灯光，虚实结合，实现现代与传统的和谐交汇。

This scenario adopts simple and fluid design techniques. It emphasizes restraint in the selection of materials and naturally expresses modern aesthetic taste through the coordination of different spatial layers. At the center of the sales department, the marble sand table with both light and dark colors and the serialized lights from the two-storey high ceiling synthesize virtuality and reality and realize a harmonious modern-traditional interaction.

久和烧肉酒场

JIUHE YAKINIKU WINERY

项目名称：久和烧肉酒场
项目设计：广东星艺装饰集团
设计师：余圣华

“隐”是通往冥想的路口。在僻静的空间中，利用水波纹和建筑水泥灰的对话，辅以灯光制造出“隐”的意境。

"Seclusion" is the intersection leading to meditation. In this secluded space, the interaction between rippling patterns and cement grey, together with light, creates a sense of seclusion.

楚雄欢乐渔港

CHUXIONG HAPPY FISHING PORT

项目名称：楚雄欢乐渔港
项目设计：广东星艺装饰集团
设计师：廖春坤

本案设计满足了时尚餐饮、多维生活、多重社交及多元活动场所的需求，灯光设计是点睛之笔。色香味俱全的菜品、空间的温度、家私的品质、色彩调性，最后由灯光交织成一个整体，营造了时尚消费的氛围。二楼采用新中式风格打造包间，在设计上延续了明清时期家居配饰理念，经典元素的现代应用，让整体风格更加雅致大气。

The place satisfies the multiple needs for fashionable food, multi-dimensional life, multiple social contacts and diverse activities. In this scenario, the lighting design is the finishing touch. Exquisite cuisines, warm space, high-quality furniture and harmonious color tone, all become a whole in interlaced light beams, creating an atmosphere for fashionable consumption. On the second floor, private rooms are in a new Chinese style. Apart from continuing the concept of home accessories of the Ming and Qing dynasties, it also integrates modern elements, rendering the overall style more elegant and more magnificent.

宏兴中心大楼

HONGXING CENTER BUILDING

项目名称：宏兴中心大楼
项目设计：广东星艺装饰集团
设计师：郑元丰

本案设计旨在将现代简约与传统中式相结合，并将人文情怀融入多元化的艺术空间。

This scenario aims to blend modern simplicity with traditional Chinese design and to integrate humanity into the diverse artistic space.

臻·境

ZHEN · JING

项目名称：臻·境
项目设计：广东星艺装饰集团
设计师：余文超

本案设计师运用现代设计手法，以抽象的形式，打造一座具有儒家文化气质的售楼中心，给人们一种惬意、简雅的生活体验。疏密有致的线条组合，宛若被串起的一页页书卷，夹杂着木质的芬芳，营造出轻盈又深厚的人文意蕴。

By adopting modern design techniques and abstract forms, the designer builds a sales department full of Confucian culture, providing a cozy, simple and elegant life experience. The dense and sparse lines look like book pages in strings and with a waft of wood scent, they create a soothing and profound humanistic atmosphere.

盘古烤肉

PANGU BARBECUE

项目名称：盘古烤肉
项目设计：广东星艺装饰集团
设计师：邱闻强

本案通过木头、装饰膜、小花砖、粗麻布等材料的运用，配以射灯为主、筒灯为辅的温和光源，使空间明暗层次柔和过渡，让人们步入空间时有温馨、舒适，甚至是慵懒的感觉。

This scenario uses materials like wood, decorative films, small patterned tiles and gunny with the spotlight as the main light source and the gentle downlight as a supplement. All these achieve mild transitions between lightness and darkness in the space. On entering it, one can feel a sense of warmth, comfort and even languor.

达州星艺办公室莲湖广场

LIANHU SQUARE, DAZHOU XINGYI OFFICE

项目名称：达州星艺办公室莲湖广场
项目设计：广东星艺装饰集团
设计师：徐珊、叶王峰

本案重点展示了空间不规则的延展营造的独特空间氛围。办公区及洽谈区有序与无序的融合，达到了至美之境。

This scenario displays a unique space created by the irregular spatial extension. The integration of the office area and the negotiation area, whether in or out of order, achieves extreme beauty.

耕艺种德
设计幸福